Alaska's Mushrooms

A PRACTICAL GUIDE

Harriette Parker

Alaska Northwest Books™
Anchorage • Seattle • Portland

DISCLAIMER: This book is not intended to be used alone as an identification guide to edible and poisonous mushrooms. It is most useful to beginners in conjunction with other, more comprehensive guidebooks (see Reading and Resources at the back of the book). Though the mushrooms described as edible in this book are widely considered safe for most people, positive identification and proper preparation of mushrooms intended for eating are the reader's responsibility. *If identity is at all questionable, don't eat any wild mushroom.* Neither the author nor the publisher is responsible for allergic or other adverse reactions individuals might experience from eating wild foods.

Second printing 1996

Library of Congress Cataloging-in-Publication Data
Parker, Harriette.
        Alaska's mushrooms : a practical guide / by Harriette Parker.
            p.   cm.
        Includes bibliographical references (p. 89) and index.
        ISBN 0-88240-453-9 (acid-free paper)
        1. Mushrooms—Alaska—Identification. 2. Mushrooms—
    Alaska—Pictorial works. I. Title
    QK617.P26 1994
    589.2'22'09798—dc20   94-6298
                                                                CIP

Photo credits: All photos are by Neil McArthur, with the exception of the following: Norma Wolf Dudiak, pp. 56, 80; Janet Klein, p. 12; John Sigler, pp. 27, 39.
Front cover: Fly Agaric, by Neil McArthur
Back cover: King Bolete, Golden Pholiota, and Gem Puffball, by Neil McArthur

Managing Editor: Ellen Harkins Wheat
Editors: Don Graydon, Carolyn Smith
Designer: Cameron Mason
Illustrator: Robert Williamson
Map: Vikki Leib

Alaska Northwest Books™
An imprint of Graphic Arts Center Publishing Company
Editorial office: 2208 NW Market Street, Suite 300, Seattle, WA 98107
Catalog and order dept.: P.O. Box 10306, Portland, OR 97210
    800-452-3032

Printed on acid- and elemental-chlorine-free recycled paper in the United States of America.

*Dedicated with much love and appreciation to my family,
Neil and Colin McArthur, whose shared enthusiasm for
peaceful hunting of wild plants, animals, and mushrooms has
fueled this book and kept it moving and fun; to my parents,
who provided a forest for a backyard and trusted their
children to the Great Spirit there; and to the outdoors and
all its offspring, from wild mushrooms to great whales.*

Acknowledgments

This project was made possible by the following people and with the assistance of many other individuals, mushroom guides, and journals not credited here. Thank you all:

Friends and members of the Alaska Mycological Society, for contributing photographs, recipes, and reports of far-flung Alaskan mushroom sightings. Peggy Blenden, co-editor of *Shroomer Rumors*. Phyllis Kempton, for her astute counsel and mycological expertise. Janice Schofield, botanist, author, and co-founder of the Alaska Mycological Society. David Arora, who has done for mushrooming what Roger Tory Peterson accomplished for bird-watching: increased by many thousands the number of people who enjoy getting outdoors to sight and record species both familiar and unusual. Biologist Meade Cadot and mycologist Rick Van De Poll, both of the faculty of Antioch New England Graduate School; and Priscilla Russell, ethnobotanist and author, for reading portions of the manuscript and offering constructive commentary. Dr. Gary Laursen, mycologist, Department of Botany and Wildlife, University of Alaska Fairbanks, for providing range reports for most of the species featured in this book. The ever-helpful, patient staff at Alaska Northwest Books: Marlene Blessing, Ellen Wheat, Don Graydon, Carolyn Smith, Stephanie Henke, and Cameron Mason.

Contents

The Mushrooms

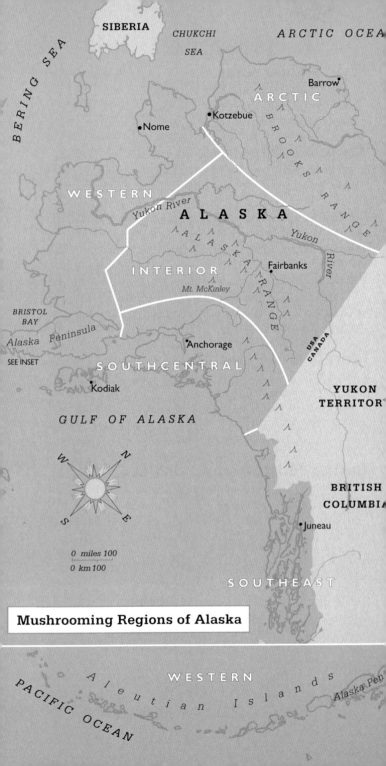

Mushrooming Regions of Alaska

Meet the Mushroom

Alaska's mushrooms are an endless source of wonder, beauty, and nourishment. This pocket guide introduces you to some mushrooms that are commonly met in Alaska and are intriguing enough to make most people wonder, "What mushroom is that?"

Mushrooms are the fleshy fruit of certain fungi. They produce and release, when ripe, millions of microscopic spores that serve to reproduce the fungi (somewhat as seeds do plants). Spread by wind and animals, fungal spores germinate if they land on suitable habitat, and produce a mass of strands that form the *mycelium,* the feeding and parent body of the fungus.

A fungus can't harness sunshine to make food as plants do, or move about to graze or capture food as animals can. Instead, the mycelium absorbs nutrients from its surroundings, and when conditions are right, forms embryonic mushrooms that swell into "buttons" and burst forth as the mushrooms that are the subject of this book. From these mushrooms, another generation of millions of spores is released to continue the cycle of mushroom creation.

Each type of mushroom-producing fungus has a preferred source of nourishment. Many get nutrition from decayed plant or animal matter; they perform a valuable service by breaking down this material and returning it to the soil. Others form a symbiotic partnership with the roots of a plant, typically a forest tree, in which nutrients and services are exchanged. A few are harmful parasites that invade the living tissue of trees to take what they need.

You can get to know mushrooms personally by observing those species appearing in your own backyard and garden or in nearby fields and woods. Many of these homebodies will prove to be lifelong neighbors. Becoming familiar with their names and habits takes only a few simple tools, the help of up-to-date regional field guides, a knowledgeable local mentor or two, and patience. For company in the wild, join the nearest mycological society and attend its outings and workshops.

Using this Book to Identify Wild Strangers

Alaska is a vast uncharted land when it comes to collecting mushrooms. There are literally hundreds of mushroom species in Alaska and few mushroom collectors. This book contains only a sampling of the best-known species found in populated areas of the state. Aspiring 'shroomers, residents, and visitors can use this book to become familiar with some of the mushrooms most often encountered and techniques for learning to identify them.

From June through September in Alaska, you're almost certain to encounter many intriguing mushrooms. If you find a mushroom you wish to identify, gather several mature specimens. Note their overall appearance, particularly the mushroom's spore-bearing surface (gills, pores, or teeth, usually found on the underside of the cap). Use one cap to prepare a sporeprint, a key clue to mushroom identity (see discussion below, About Sporeprints). Read the brief introduction to each of the seven mushroom groups included in this book until you find the category that most closely matches the mushroom you are trying to identify. Thumb through the photos. Does your mushroom look like one of these? Carefully compare all the entries in that section with your mystery mushroom.

If the mushroom you have found has *all* the features described in one of the entries, you've probably made a correct identification. If not, consult other current field guides, several of which are listed in Reading and Resources at the back of the book, to help you identify your mushroom. Any mushroom new to you that you intend to eat should be identified in more than one current guide and verified by a local knowledgeable expert.

Each entry in this book includes the time of year and habitats in which the mushroom grows and names of look-alikes. Range information is provided, indicating the region of Alaska in which a particular species is *reported* to be found. Because Alaska is so large and the number of 'shroomers so small, unreported species (such as the deadly *Amanitas*) might well occur here. Finally, the entry tells you whether the mushroom is considered poisonous or edible and provides specific cautions needed for your safety in collecting or eating that species.

Mushrooms are described in this book simply as small, medium, or large, based on cap width. Yet you might run across a specimen that is smaller or larger than its expected size because of variations in such factors as rainfall, nutrients, temperature, and heredity. You'll develop a feel for size with experience. I don't know anyone who packs a tape measure when collecting, though some serious mycologists must! Some mushrooms, like Shaggy Manes, can be medium in cap width but appear large due to stalk height. The size descriptions in the text translate into the following measurements:

> Small—½ inch to 2 inches wide
> Medium-size—2 to 6 inches wide
> Large—6 inches wide or wider

Most collectors note freshness-dependent features like color and odor while they are in the field, then hasten home with their booty to complete the remaining detective work in relative comfort.

Mushrooms are best collected for identification when newly mature. The spores are ripe and ready to drop, and sporeprints are more reliably taken then. Very immature or overmature mushrooms don't produce spores.

Hunting Mushrooms in the Wild

Like most people, I began my collecting career by tearing into the woods, yanking up 'shrooms with my bare hands, dumping them in grocery bags, and moving on. Now I know better.

You'll want to have the right tools and supplies for your expedition. Here is a list of the few things you need and how to use them:

The author and her son holding Golden Pholiota (*Phaeolepiota aurea*), also known as Alaskan Gold.

- Trowel or knife to dig the entire mushroom from the soil. It's important to collect not just the stalk and the cap but also the stalk base, which is often buried in the soil. The base can be important in identifying the mushroom. Cuplike or sacklike tissue surrounding the base is characteristic of some deadly poisonous *Amanita* species.
- Soft toothbrush to remove soil, debris, and insects from mushrooms.
- Waxed-paper bags or a roll of waxed paper for holding the mushrooms you collect and keeping them fresh. (Plastic bags aren't good because they cause mushrooms to sweat and deteriorate quickly.) Bag each kind of mushroom separately so it will be easier to identify later. More important, keep different species separate to help prevent mixing of poisonous varieties with the ones you plan to eat for dinner.
- White note cards for taking sporeprints in the field.
- Basket for storing your finds.
- Hand lens (10x) for seeing tiny features.
- Current regional field guide.
- Small pad for jotting down field notes. Mushrooms tend to show up each year at roughly the same time and place, and good notes can direct you to favorite spots

each season. More serious 'shroomers should keep a notebook to record useful information, including the name of the mushroom, date, location, what it was growing on, and anything else you would like to remember. I also note the appearance of "new" or old mushrooms each season on my calendar at home.

About Sporeprints

Sporeprints reveal the color of a mushroom's spores, providing an essential clue to the mushroom's identity. Many serious mushroom poisonings could be avoided if collectors would routinely determine the spore color of mushrooms they intend to eat. Though most often useful in identifying gilled mushrooms, sporeprints can also be made from boletes, teeth, and coral mushrooms.

To make a sporeprint, cut the cap off a mature mushroom and lay it with the gills down on a piece of white paper. (With non-gilled mushrooms, put the spore-bearing surface down.)

Cover the cap with a bowl to prevent air currents from disturbing the spores. (Covering the cap also protects people in your household from inhaling harmful spores.) Enough microscopic spores will usually drop and collect on the white paper over the next two to twelve hours so that you can make out their color. (To see white spores on white paper, look at them at an angle. Black paper is sometimes used, but it distorts color perception.) If mushrooms are too old, too young, or are dry, soggy, or sterile, a print can't be taken.

You can also start sporeprints in the field. Place the mushroom cap, gills down, on a white note card, wrap it in waxed paper, and arrange it carefully in the bottom of your basket. You may have a sporeprint by the time you arrive home.

Sporeprints sometimes occur naturally on a mushroom's stalk, on other mushrooms, or on the ground below the cap. But these wild sporeprints can be misleading. It's safest to collect the sporeprint on white paper, especially when you're looking for mushrooms to eat.

Safe 'Shrooming

The first question most people ask about a mushroom is, "Can I eat it?" The answer—for most people and for many mushrooms—is a qualified "yes." But it isn't a decision to be made lightly. Following are some important safety guidelines.

- Be as certain as possible that you've correctly identified a mushroom species before eating it. Consult several current regional field guides and ask a knowledgeable local collector to examine and identify your collection. When in doubt, throw it out!
- Learn to recognize poisonous look-alikes of species you collect for the table.
- Keep different species in separate containers when you collect unfamiliar mushrooms so that poisonous and edible pieces won't be mixed.
- Collect only healthy, insect-free, mature mushrooms for the table. Avoid very young and old specimens.
- Collect mushrooms in areas uncontaminated by pollutants (vehicle exhaust, herbicides, etc.).
- Cook all wild mushrooms thoroughly.
- Sample only a tiny portion (one tablespoon) of any edible wild mushroom new to you, as you might suffer

A basketful of puffballs.

an allergic reaction. If no symptoms develop within a couple of days, you are probably not allergic.

- Eat modest servings of edible wild mushrooms, even those familiar to you, to avoid indigestion.
- Avoid drinking alcohol when eating a species for the first time. Alcohol sometimes combines with chemicals in mushrooms to create unpleasant side effects in some individuals. Some morels and *Coprinus* species are examples.
- Avoid serving wild mushrooms to young children or to people with impaired health.
- Guard young children to keep them from eating mushrooms in the wild.
- Avoid inhaling spores, which may cause respiratory disorders in some people.
- Avoid panic if you become ill after eating mushrooms. Purge your stomach/intestinal tract, and consult a doctor at once. Show the doctor an uncooked specimen of the mushroom you ate.

Collecting with Care

Whether you hunt mushrooms for food or for fun, you can help protect Alaska's mushrooms by keeping the woodlands, fields, and other places where mushrooms grow in your community clean, healthy, and undisturbed.

- Never rake the forest floor to uncover the mycelium (the underground body of the fungus) and hidden "buttons," and never dig up more mushrooms than needed for food or identification purposes. This can damage the mycelium and destroy the mushroom patch.
- Collect mushrooms sparingly, only for your personal needs. Leave a few to release spores, renew the resource, and delight other seekers.

The Questions I Hear Most

Where do mushrooms grow in Alaska? And when?
Mushrooms grow wherever temperatures and rainfall permit, which means almost all of Alaska as far north as the

Arctic Circle and a bit beyond. The fruiting season over most of the state ranges from spring snowmelt (April through June) until fall freeze (September through November).

What are the most dangerous kinds of mushrooms in Alaska?
Potentially fatal mushrooms, often spotted in populated areas, include the Fly Agaric *(Amanita muscaria)*, Poison Pax *(Paxillus involutus)*, Sulfur Tuft *(Naematoloma fasciculare)*, false morels such as the Brain Mushroom *(Gyromitra esculenta)* and the Hooded False Morel *(Gyromitra infula)*, *Cortinarius* species, and *Galerina* species. And there are almost surely more out there; Alaska has a huge outdoors with only a handful of people looking for mushrooms. (This book includes photos and descriptions of the Fly Agaric, Poison Pax, Brain Mushroom, Hooded False Morel, and one *Cortinarius* species, *Cortinarius semisanguineus.)* To date, there have been no reported fatalities due to mushroom poisoning in Alaska.

Which mushrooms in Alaska are probably the safest for beginners?
Hedgehogs *(Hydnum repandum)* and Shaggy Manes *(Coprinus comatus)*—both in this book. They're easy to identify and, for most people, they're easy to digest.

How many kinds of mushrooms do you eat?
In Alaska, about eight. In spring, Black Morels (*Morchella* species). In summer, King Boletes *(Boletus edulis)*, Shaggy Manes *(Coprinus comatus)*, Meadow Mushrooms *(Agaricus campestris)*, and the rarely found Prince *(Agaricus augustus)*. In fall, Hedgehogs *(Hydnum repandum)* and Combs *(Hericium ramosum)* or other *Hericium* species should I chance to catch them fresh, and Tumbleballs *(Bovista plumbea)*. (This book includes individual entries for all these mushrooms except the Prince.)

What is your favorite mushroom?
For beauty: Golden Pholiota. For fun: morels; the thrill of the hunt! For eating: Hedgehog or King Bolete. Don't make me choose! Best all-around: Shaggy Mane.

14

Gilled Mushrooms

G illed mushrooms, also called agarics, have a fleshy cap and stalk (or sometimes a cap only). Their spores are borne on radiating bladelike gills beneath the cap. Gilled mushrooms are usually found on the ground or on wood, in forests or open areas. More than 2,000 gilled mushroom species have been reported in North America, including some that are choice edibles and others that are dangerous poisoners.

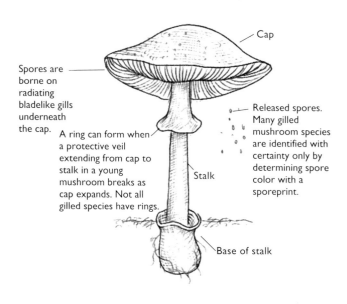

Cap

Spores are borne on radiating bladelike gills underneath the cap.

A ring can form when a protective veil extending from cap to stalk in a young mushroom breaks as cap expands. Not all gilled species have rings.

Released spores. Many gilled mushroom species are identified with certainty only by determining spore color with a sporeprint.

Stalk

Base of stalk

Fly Agaric
POISONOUS
Amanita muscaria (Family: Amanitaceae)

A normally imperturbable innkeeper called me in some excitement one summer day to report the sudden appearance of a single fabulous red-and-white fungus on his lawn. Could I come and identify it?

No need! This could be none other than *Amanita muscaria*, mushroom of a thousand storybook illustrations.

It looks like a mushroom that has stories to tell, and it is. Records of its use as an intoxicant or hallucinogen for recreational or religious purposes go back centuries in parts of Northern Europe and Asia and among Native American tribes. People in parts of Europe mashed it into milk for use as an insecticide (thus the name "Fly Agaric," or fly fungus).

"How can I preserve it?" the innkeeper asked, wanting to save its beauty. "You can't," I replied, "except in pictures. Why don't you get out your camera and just preserve it on film?"

Mushroom authority David Arora suggests that the Fly Agaric is a frequent neighbor of the King Bolete, a much-sought edible. Sure enough, I finally found this prize next to a "fly" patch in spruce/birch/alder brush along a busy road in Homer.

Also known as: Fly Amanita.
Where to find: On the ground, usually beneath or near spruce, birch, or mixed woods. Reported from Southeast, Southcentral, and Interior Alaska.
When to find: Summer and fall.
Look carefully for all these features:
- Medium-size to large mushroom.
- Cap spattered with whitish wartlike tissue. Color usually red, but can vary to orange, yellow, or brownish shades. White forms produced by weather, sun exposure, or nutritional or genetic factors are found on rare occasions.
- Flesh white.

- Gills white, closely spaced, free from stalk.
- Stalk white, with a droopy ring near top.
- Base of stalk enlarged, circled with bands of fluffy white scales.
- Sporeprint white; spores borne on gills under cap.

Look-alikes: The white form resembles other white species of *Amanita* and some large *Lepiota* and *Agaricus* species.

Warning: Poisonous. Eating this mushroom can produce varying unpredictable physical reactions from mild giddiness, alcohol-like intoxication, and deep sleep to delirium and convulsions. Occasional deaths occur in susceptible adults or young children. Because the genus *Amanita* can be diffi-cult to identify and several of its species contain deadly toxins, I advise admiring them all as you would the stark scarlet presence of baneberries on your path or the volcanic plume over Cook Inlet from Mount Augustine. Nod respect for their beauty and move on.

Meadow Mushroom
Agaricus campestris (Family: Agaricaceae)

The Meadow Mushroom was the first wild mushroom I ever tasted. A neighbor brought over a basketful of the fresh beauties from a pasture by the Bay of Fundy in eastern Canada. They were consumed in a matter of minutes.

The next morning I set off early to gather my own bagful while the cattle lowed balefully. Then I checked out the single mushroom guide in the county library bus.

Back home I fried and ate at least a dozen mushrooms before settling in to read about the painful fate met by collectors of "Pinkbottoms" who carelessly gathered poisonous Meadow Mushroom look-alikes. I lost my appetite for a while.

Later that season I also discovered the alluring *Amanita verna* (known as the Destroying Angel) along my cabin trail. I thanked my lucky stars for good timing, and I buried the deadly angels in a briar patch. This dangerous look-alike and others from the *Amanita* gang could be mistaken for their *Agaricus* doubles, with possibly fatal consequences. Most people who take care to get to know the Meadow Mushroom can eat this savory mushroom safely.

Soon after arriving in Alaska, I discovered my first fungal love again—the Meadow Mushroom, old Pinkbottom—waiting for me in a neighbor's pasture.

Also known as: *Psalliota campestris*, Field Mushroom, Pinkbottom, *Champignon*.

Where to find: Often growing in groups, in the grass of pastures, lawns, and meadows (not in forests). Reported from Southcentral and Interior Alaska.

When to find: Summer and fall.

Look carefully for all these features:

- Small to medium-size mushroom.
- Cap white, dry, silky.
- Flesh white, thick, doesn't change color when rubbed or broken.
- Gills bright pink, aging to blackish brown; crowded; free from stalk.
- Veil covers gills at first, sometimes leaving a fragile ring on stalk.
- Stalk straight, white, short, thick; usually tapered toward base.
- Stalk base (bottom) is not swollen, not enclosed in a cup or sacklike tissue, not circled with coarse scales, and does not turn yellow when bruised.
- Odor mild, neither antiseptic nor sweet (you can smell the crushed stalk base).
- Sporeprint dark brown; spores borne on gills under cap.

Look-alikes: *Amanita* species (some deadly) and *Lepiota* species (some poisonous); other *Agaricus* species (including some sickeners).

Food use: Edible with caution. Almost everyone recognizes the Meadow Mushroom at a glance as glorified kin of the familiar grocery-store mushroom. However, even edible *Agaricus* mushrooms (including the grocery-store variety) cause mild to severe stomach distress in a good number of consumers, and wild *Agaricus* can be tricky to identify. Heeding the following cautions can provide a safe start to your future with Pinkbottoms.

Caution: Every few years in North America someone becomes seriously ill or dies because of mistaking deadly *Amanita* look-alikes for popular edible *Agaricus* mushrooms, and severe gastrointestinal disturbances result from eating poisonous look-alike *Agaricus* or *Lepiota* mushrooms. To be safe when beginning to collect the Meadow Mushroom, check all the identifying features listed here with your specimen in hand, paying special attention to sporeprint and stalk base. Then show the mushroom to a friendly local expert to confirm your identification before heating your skillet.

Orange Delicious
Lactarius deliciosus (Family: Russulaceae)

The Orange Delicious inhabits with special vigor its immense spruce bog territory in Alaska. You can bet your basket that if you wander into a moist spruce stand in autumn you'll strike gold—unless claim jumpers get there first!

These edible mushrooms are a favorite of beginning 'shroomers because they're especially easy to find and recognize as a group. Just don't expect them all to fit in one mold. The Orange Delicious are individuals, varying in size and in proportions of orange and green coloration. They can secrete either abundant or scant latex (the bright orange fluid that oozes from the mushroom when broken). And they might offer you either a pleasant or, despite their name, a bitter taste.

This commonly met but always striking fungus came to human attention early. A mural depicting Orange Delicious was found in the ruins of Pompeii, buried almost 2,000 years ago by the eruption of Mount Vesuvius.

Also known as: Saffron Milkcap.
Where to find: Growing scattered or in groups, on ground or moss among conifers in moist forests, parks, or lawns. Reported from Southeast, Southcentral, and Interior Alaska.
When to find: August through September.
Look carefully for all these features:
• Small to medium-size mushroom.

- Cap orange with green tints, sometimes brown or grayish hued; smooth.
- Flesh orange, brittle.
- A bright orange fluid called latex is usually secreted when the mushroom is broken; latex can most easily be seen where gills and flesh meet.
- Gills orange, attached to stalk or running down stalk a bit.
- Stalk orange and brittle, short and stout.
- All parts turn green with age or handling.
- Sporeprint pale yellow; spores borne on gills under cap.

Look-alikes: Similarly colored *Lactarius* species. My students have mistaken unusually golden-tinted old Poison Pax for Orange Delicious.

Harvesting tip: Avoid collecting specimens that are predominantly green-colored. This might indicate the mushroom is too old and bitter for table use.

Food use: Edible, with caution. Opinions on the taste range from "ahhh" to "blah." A crumbly texture, often bitter taste, and moldy green tinges discourage many would-be diners. I've avoided these complaints by collecting only young, moist, orange-all-over specimens for the table, sautéing them low and slow, then adding them to cooked pasta in a covered casserole.

Caution: Collect young, predominantly orange-colored mushrooms and cook thoroughly. Examine each specimen to avoid confusion with Poison Pax or other possibly toxic or untasty mushrooms. Fry up several samples. If they taste bitter after thorough cooking, toss them out.

Red-hot Milkcap
POISONOUS
Lactarius rufus (Family: Russulaceae)

The Red-hot Milkcap and the mushroom called the Sickener (see the next entry) are nauseators that lure foragers, especially young children, with attractive candy colors and texture.

My son began berrying and mushrooming as an observer from his dad's backpack at the age of six months. By age three he had survived two baneberry incidents and one encounter with the Red-hot Milkcap. The milkcap he somehow picked and popped into his mouth along a trail just a few steps in front of me. I noticed when he began screaming, "My mouf, my mouf!"

"What did you put in?"

"A mushwoom!"

Just as I began to panic I saw some mashed reddish brown glop on his jacket. "Is this what you ate?"

"Yes, *Actaweus wufus*!"

I rinsed his mouth out. There were no further symptoms, so I guessed he hadn't swallowed any. He hasn't seemed tempted to sample any wild food save raspberries since that scare.

I had a call once regarding a two-year-old girl who took a bite of something that sounded like Red-hot Milkcap. She too experienced no serious symptoms. But the Red-hot Milkcap is potentially dangerous because the initial taste sensation is mild, and a person could eat an entire cap before the peppery alarm rings. Especially eaten raw, the Red-hot Milkcap and the Sickener have been reported to cause severe gastrointestinal disturbances.

These mushrooms are abundant and easily spotted. The name, Red-hot Milkcap, refers to its taste and color and to the milklike fluid, often termed latex, that oozes from cap tissue.

Where to find: Growing scattered or in groups, on ground or moss in moist or boggy conifer forests or mixed

22

woods. Reported from Southeast, Southcentral, Interior, and Arctic Alaska.

When to find: Late summer and fall.

Look carefully for all these features:

- Medium-size to large mushroom.
- All parts of the mushroom are reddish brown and brittle.
- Cap dry, smooth, with tiny knob in center.
- When mushroom is broken, a white milklike fluid called latex is secreted and is especially apparent where gills and flesh meet.
- Gills pale, tinged cap-color.
- Taste is bland, gradually becoming peppery hot. (Caution: If you *must* taste this mushroom, chew a tiny piece and spit it out as soon as you sense a peppery taste.)
- Sporeprint white; spores borne on gills under cap.

Look-alikes: Other similarly colored, related *Lactarius* and *Russula* mushrooms. Sometimes mistaken for the mushroom known as the Deceiver, *Laccaria laccata*, which can appear close in color and occurs in the same places.

Warning: Poisonous. In North America it is widely reported that eating this mushroom can cause gastric disturbance, occasionally severe. In Europe the (apparently) same Red-hot Milkcap and other similar peppery members of the *Russulaceae* family are canned and used for flavoring. This might be an example of one kind of mushroom developing different qualities in different places. Until the facts are settled, it's advisable to just pass by this offering from the outdoor grocery store.

Sickener
POISONOUS
Russula emetica (Family: Russulaceae)

This clean-cut candy cane of a mushroom imparts a Christmasy feeling throughout the areas of young spruce and sphagnum moss where it thrives. It's usually the first mushroom a person spots when entering the woods, and it's always a downer to introduce children, or adults, to the Sickener—food for the eyes only! Its scientific name is perfectly descriptive: in Latin, *Russula* means red and *emetica* means that it induces vomiting.

Where to find: Growing scattered or in small groups, on ground or moss in moist or boggy spruce forests or mixed woods. Reported from Southeast, Southcentral, Interior, and Arctic Alaska.
When to find: Late summer and fall.

Look carefully for all these features:
- Medium-size mushroom.
- Cap bright red, fading with age.
- Flesh white except for pink tinge directly beneath cap skin.
- Gills white.
- Stalk white, snaps like chalk when broken.
- All parts of mushroom brittle.
- Taste is instantly intensely irritating. (Caution: If you *must* taste this mushroom, chew a tiny bit of cap and spit it out immediately.)
- Sporeprint white; spores borne on gills under cap.

Look-alikes: Other red-and-white *Russula* species.

Warning: This mushroom can cause vomiting and nausea.

Sheathed Waxy Cap
Hygrophorus olivaceoalbus
(Family: Hygrophoraceae)

Woods that are a snarly morass of dead ends all summer turn open and inviting come September when the Sheathed Waxy Caps take to the floor in a brief appearance before fall ends.

Take time to catch their class act. These mushrooms

look like petite penguins parading through the trees in their spiffy dark-and-white uniforms.

The waxy caps don't make for fine dining. They can, however, be an ingredient in a superb tonic for tired spirits: just take a 20-minute stroll through a forest full of them.

Where to find: Often growing abundantly in scattered small groups or singly, on soil or moss in spruce or other conifer woods. Reported from Southcentral Alaska.
When to find: Fall, sometimes surviving a frost or two.
Look carefully for all these features:
- Small to medium-size mushroom.
- Cap olive, turning dark brownish black toward center, streaked with small blackish fibers, often slimy.
- Flesh white, thick, soft.
- Gills white, waxy, soft, often descending stalk slightly.
- Stalk white and smooth just above a flaring ring; sheathed with slimy fibrous dark and light bands below the ring to base.
- Sporeprint white; spores borne on gills under cap.
Look-alikes: Other slimy, similarly colored and patterned waxy caps.
Food use: Not recommended. The few published sources that include this waxy cap describe it as a slimy take-it-or-leave-it mushroom at best.

GILLED MUSHROOMS

Deceiver
Laccaria laccata
(Family: Tricholomataceae)

This mushroom isn't called the Deceiver for nothing. It appears in a mind-boggling variety of colors, shapes, and sizes, and can move into almost any old place and call it home. Birders who can identify a sparrow when it's only a dot on the horizon often need a lot of exposure before unmasking the Deceiver! Yet someday, if you go in among the spruce enough times, there the Deceiver will be, clicked forever into focus for you.

Deceivers were the first wild mushrooms I identified and decided to eat on my own. Since then I've come to appreciate them more for their visual appeal than for their so-so taste. When the sun lights up their lustrous pinks, oranges, reds, and browns, I never resist picking a couple to sniff their fresh soapy smell and admire their beauty.

Where to find: Growing scattered or grouped, in open places on ground in moist conifer forests or mixed woods. Reported from Southeast, Southcentral, Interior, and Arctic Alaska.

When to find: During cool spells from late summer into fall. These mushrooms resist decay and can stand for weeks in the woods looking as good as new.

Look carefully for all these features:
- Small to medium-size mushroom.
- Cap pinkish, orangish, reddish, or brownish; has tiny soft scales and a small dimple or hole in the center.
- Flesh cap-colored, thin, soft.
- Gills pink or orange, waxy, spaced far apart, might run down stalk a bit.
- Stalk cap-colored, tough, fibrous, sometimes twisted.
- Odor soapy.
- Sporeprint white; spores borne on gills under cap.

Look-alikes: Other *Laccaria* species and some same-size, pinkish-gilled mushrooms, a few of which are poisonous.

Food use: Edible, with caution. Edibility ratings from many sources range from "Why bother?" to "Pretty good if seasoned enough." I ate the Deceiver regularly in pasta dishes when I began 'shrooming, but soon moved on to tastier types.

Caution: The Deceiver is often difficult to identify and can be confused with small poisonous mushrooms. I advise beginners to practice finding this fungus for a while before dining on it. These are good candidates for running by your local experts or checking out in more than one field guide.

Fried Chicken Mushroom
Lyophyllum decastes
(Family: Tricholomataceae)

To eat or not to eat? The temptingly named Fried Chicken Mushroom epitomizes this occupational dilemma for 'shroomers. Here's a mushroom you can hardly walk away from. Few can resist grabbing and bagging the juicy clumps springing up so abundantly and conveniently

along sidewalks and on lawns at about the time other fungi have packed it in for the season.

So what's the rub? Unfortunately, the only way to know what you've got—be it good, bland, bitter, or nauseating—is to dive in and taste it. It's just such wishy-washy mushrooms as these that cause some of us to hunt mushrooms for sport rather than food!

Also known as: *Clitocybe multiceps.*

Where to find: Growing in dense clumps, on ground along sidewalks and driveways, on lawns, on buried woody debris, in gardens and other disturbed ground. Reported from Southeast, Southcentral, and Interior Alaska.

When to find: Late summer and fall.

Look carefully for all these features:
- Medium-size mushroom, appearing in large, dense clumps with stalks joined at base.
- Cap usually brownish; feels moist, smooth; wavy-edged due to crowding.
- Flesh white, thin, firm.
- Gills white, attached to stalk.
- Stalk white, tinged brownish, usually bent out of shape by crowded growth.
- Sporeprint white; spores borne on gills under cap.

Look-alikes: Other clustered mushrooms with joined stalks, some poisonous, like the pink-spored *Entolomas* or the white-spored *Clitocybe dilatata.*

Food use: Questionable. The mushroom got its name because someone thought that, salted and fried in butter, it approached the flavor and crunchiness of fried chicken. But reports of stomach upset involving Fried Chicken Mushrooms are not uncommon. Opinions on the tastiness of this quick larder-filler run from thumbs up to thumbs down, with most in between. Things haven't changed much since the turn of the century, when New York botanist Charles Peck noted that "Some pronounce it among the best mushrooms when cooked; others say it's unfit to eat."

Caution: If you can't resist Fried Chicken Mushrooms, first pay special attention to careful identification to be sure you have the right species. Then sample only a very small portion of well-cooked cap at a sitting.

Late Oyster Mushroom
Panellus serotinus
(Family: Tricholomataceae)

Some chilly day in late autumn, with mushrooms long gone and forgotten, you'll come across a stump or log freshly capped with a handsome harbinger of winter: the Late Oyster Mushroom.

Though fungi growing on dead or dying wood are often loathed as dangerous forest parasites, the Late Oyster Mushroom is described as a "weak parasite." It invades trees already stricken by disease or injury, sort of like a lame wolf attacking a sick moose. When the host tree expires, the fungus then carries out the beneficial work of breaking down the dead tree tissue and returning it to the soil.

Also known as: *Pleurotus serotinus*, Greenback.

Where to find: Growing scattered or in overlapping clusters, usually on dead or dying hardwoods, sometimes on conifers. Reported from Southcentral and Interior Alaska.

When to find: Fall.

Look carefully for all these features:
- Small to medium-size mushroom.
- Cap colored green, yellow, brown, or violet; smooth; oyster-shaped.
- Flesh cream-colored, thick, firm.
- Gills yellowish orange.
- Stalk, when present, is a yellowish, hairy short stub connecting cap to host tree.
- Sporeprint cream-colored; spores borne on gills under cap.

Look-alikes: Other *Panellus* species; *Panus* and *Crepidotus* species.

Harvesting tip: Use a knife to detach caps from the tough stubs attaching them to the tree.

Food use: Edible, with caution. One guidebook describes this mushroom as tender and flavorful if cooked long over low heat, while another says it's inedible. Most conclude it's

a mediocre mushroom, too bitter to bother collecting for food. Everyone gives it credit for being available when there's nothing else. Collecting young fresh caps before frost sets in and cooking them low and slow might help the ratings. Or perhaps these stalwart specimens are to be appreciated most for providing that last muted surprise of color before the woods shut down for the long winter night.

Caution: Late-season mushrooms are sometimes exposed repeatedly to freezing and thawing conditions, and such exposure can produce toxins.

Angel Wings
Pleurocybella porrigens
(Family: Tricholomataceae)

Finding Angel Wings usually takes more than beginner's luck. There are several devilish look-alikes, inedible and little-known, that can confuse matters.

Simply follow a seasoned 'shroomer some September into almost any conifer community in Alaska or elsewhere in North America. Before long your guide will pause beside

the trail. There, on the rotting remains of a spruce giant, the fine porcelainlike aura of Angel Wings will appear, and you will never again confuse it with anything else.

Also known as: *Pleurotus porrigens, Pleurotellus porrigens.*
Where to find: Growing scattered or densely in overlapping clusters, in woods on very decayed spruce stumps or logs, often growing out through moss. Reported from Southcentral Alaska.
When to find: Late summer and fall.
Look carefully for all these features:
- Small to medium-size mushroom.
- Cap white, delicate; shaped like a fan, clam, or shelf.
- Flesh white, thin, pliant.
- Gills white with smooth edges, crowded.
- Stalk attachment stubby or absent.

- Sporeprint white; spores borne on gills under cap.

Look-alikes: The edible and related Oyster Mushroom, *Pleurotus ostreatus,* and other *Pleurotus* species. Also a few possibly poisonous, small to large, whitish brown-spored *Crepidotus* species, growing on trees or woody debris.

Harvesting tip: Use a knife to separate the cap from its log or stump.

Food use: Mediocre for eating; bland-tasting and fragile. This flimsy little fungus works well if it's first sautéed briefly over low heat with butter or oil (seasoned with garlic, basil, marjoram, or some such), then added to pasta sauce, gravy, or omelettes.

GILLED MUSHROOMS

Deer Mushroom
Pluteus cervinus (Family: Pluteaceae)

This quietly appealing mushroom hidden against its woody perch in a cloak of brownish-tan isn't unlike a shy young deer, suddenly still, camouflaged in the forest.

Also known as: *Pluteus atricapillus,* Fawn Mushroom.

Where to find: Growing alone or in small groups, on decaying wood such as stumps or logs and on the ground over sawdust and other buried woody debris; in our community of mixed woods, they appear to prefer hardwoods. Reported from Southcentral and Interior Alaska.

When to find: Late summer to fall freeze.

Look carefully for all these features:
- Medium-size mushroom.
- Cap light to dark brown, sometimes streaked with black-brown fibers.
- Flesh white, soft.
- Gills white, aging pink; crowded, soft, free from stalk.
- Stalk white, often streaked with brown fibers, solid; easily detached from cap.
- Sporeprint pink-salmon; spores borne on gills under cap.

Look-alikes: Other *Pluteus* species and pink-spored *Entoloma* species.

Food use: Not recommended. Though edible, the Deer Mushroom is hardly a gourmet delight. Its flavor isn't consistently good, and its texture tends toward mushy.

Caution: Take care not to confuse Deer Mushrooms with the poisonous, pink-spored *Entoloma* species. For food, collect only those pink-spored mushrooms that grow on stumps or logs and surely fit each feature in the species description above, including gills that are, without doubt, free from the stalk.

Inky Cap
Coprinus atramentarius
(Family: Coprinaceae)

The first memorable mushroom sighting for many people is a small gray clump of caps that appears one day along a sidewalk or driveway. The next day, they discover in amazement, all that remains is an inky puddle where the clump had been. This is the Inky Cap, a name that refers to this mushroom's trick of dissolving into a black fluid.

And for a select few, these common Inky Caps will provide the first rite of passage on their quest to join the bold-about-mold, fearless feeders on feral fungi: the elite company of 'shroomers 'round the table! But before you drink to your good fortune, be sure to take a long look at the unusual caution at the end of this entry.

Where to find: Growing in clusters or groups, on ground with buried woody debris; on the edge of roads, parking lots, and forests; in lawns and gardens. Reported from Southeast, Southcentral, and Interior Alaska.
When to find: Late summer and fall.
Look carefully for all these features:
- Small to medium-size mushroom.
- Cap gray-brown, turning inky with age; bell-shaped or

conical; margin becoming tattered and inky with age.
- Flesh thin, colored like gills.
- Gills white or grayish, turning inky with age; crowded close together; free of the stalk.
- Stalk hollow, thin, with hairy ring sometimes visible.
- Sporeprint black; spores borne on gills under cap.

Look-alikes: Other *Coprinus* species.
Food use: Edible, with caution, for nondrinkers only.
Caution: Do not drink alcohol a few days before, while, or within a week of eating Inky Caps. This mushroom contains a toxin that in combination with alcohol causes severe reactions, including breathing difficulty, heart and blood pressure irregularities, headache, and vomiting.

GILLED MUSHROOMS

Shaggy Mane
Coprinus comatus (Family: Coprinaceae)

I can't think of a more user-friendly fungus. "Shags" are not only easy to identify, they also abound in season and are considered edible for most people.

Shags parade along streets and congregate around parking lots in downtown Homer. The largest Shaggy Mane I've ever seen grew in a well-kept lawn on the University of Alaska campus in Fairbanks in mid-September. The tall cylindrical cap was 13 inches high—and this is not a frontier tall tale!

Pioneers once made a very passable ink by boiling mature Shaggy Manes or Inky Caps, straining the fluid, and adding an anti-mold ingredient.

Also known as: Lawyer's Wig, Horsetail Mushroom, Shags, and Tennis Court Bane (because of their tendency to settle beneath pavement and then break through).
Where to find: Along roadsides; in fields and lawns; in gardens, compost heaps, and other disturbed ground; common in populated areas. Reported from Southeast, Southcentral, Interior, and Arctic Alaska.
When to find: Summer and fall.

Look carefully for all these features:
- Medium-size to large mushroom.
- Cap white, cylindrical, with brownish-ochre crown, covered with brown and white scales.
- Cap edge presses against stalk, then curls out and up as cap begins to age and to darken, eventually dissolving into a black inky liquid.
- Gills change color from white to pink and then darken to black as mushroom matures; gills packed closely together like pages in a book.
- Stalk long, white, and hollow, circled by a loose ring where stalk meets cap edge. Enlarged base is visible when collected.
- Sporeprint black; spores borne on gills under cap.

Look-alikes: Resembles *Lepiota*, *Agaricus*, *Amanita*, and other *Coprinus* species, a few of which are sickeners.

Harvesting tips: Shaggy Manes can be collected in mild spells after sufficient rain during summer and fall. For best flavor and consistency, and because this fungus deteriorates rapidly, collect specimens before their gills blacken. Gently brush or wipe these fragile mushrooms before placing them in a container. Most gatherers discard the stringy stems.

Food use: Edible and tasty. Shaggy Manes are one of the best-known edibles and are considered one of the safest. Cook or preserve caps as soon as possible to arrest the maturation process during which the mushroom's white gills rapidly darken and dissolve into something resembling an oil slick. Because Shaggy Manes tend to be watery, I sauté them in a little oil and butter, pouring off the liquid when the caps become firm and saving the liquid for flavoring soup or sauce. The taste and texture of young Shags have been likened to that of asparagus or octopus.

Caution: A few incidents of unpleasant symptoms have been reported from drinking alcohol while consuming Shaggy Manes (similar to the symptoms associated with Inky Caps; see separate entry). Young Shaggy Manes in their button stage should be avoided when they are found apart from mature mushrooms because they can be confused with buttons of poisonous *Amanita, Lepiota,* and *Agaricus* species. Overmature black-gilled Shaggy Manes occasionally cause indigestion. Avoid collecting roadside Shags, as they could be contaminated from traffic exhaust or herbicides.

Dung Dome
Stropharia semiglobata
(Family: Strophariaceae)

Here's an eye-catching little fungus with little appeal for human foragers. The Dung Dome's life-cycle probably begins with a horse, moose, or other grazer eating grass scattered with spores released by the mushroom. The spores remain in the animal's excreted feces, germinate, and develop into shiny yellow-capped mushrooms.

Other spores alight on freshly manured fields or compost, and these spores might germinate, skipping the intestinal tour.

The common name Dung Dome, proposed by mushroom authority David Arora, is not yet in wide use. But it's catchy and descriptive, and apt to take hold.

Where to find: Growing alone or in small groups, on animal dung, in manured fields, pastures, compost heaps. Reported from Southeast, Southcentral, and Interior Alaska.
When to find: Summer and fall.
Look carefully for all these features:
- Small, slimy mushroom.
- Cap smooth, yellow, and rounded.
- Flesh yellowish, thin, soft, doesn't stain blue.
- Gills gray, aging to black, attached to stalk.

- Stalk yellowish white, long and thin, often enlarged at base.
- Fragile ring (veil remnant) often circles stalk, becoming spore-blackened with age.
- Sporeprint purple-brown to black; spores borne on gills under cap.

Look-alikes: Other small dung-inhabiting *Stropharia*, *Psilocybe*, and *Panaeolus* mushrooms might be mistaken for Dung Dome.

Food use: Not recommended. Generally considered too small, slimy, mediocre-tasting, and "down and dirty."

No Common Name
(EDIBILITY UNKNOWN, POSSIBLY DANGEROUSLY POISONOUS)
Cortinarius semisanguineus
(Family: Cortinariaceae)

Mushrooms, like people, need to achieve widespread esteem or notoriety to gain a commonly used nickname. This shy, little-known mushroom has not yet earned a truly common name that I'm aware of. I sometimes call it "Red-gilled Yellow Cort" to help students identify it.

I'll never forget meeting this species during my first autumn among Alaska's mushrooms. I was absentmindedly turning over a small yellowish brown mushroom to check out the gills. Suddenly my eyes grew wide as I quite literally experienced breathtaking beauty. The gills, of a splendid scarlet hue, seemed to transform this drab little 'shroom into a golden chalice of rubies. Racing home to my mushroom books, I learned its name, *Cortinarius semisanguineus*.

Every fall these mushrooms are featured in a nearby spruce-sphagnum exhibition. We delight in showing off the mushrooms to visitors, much the same as New Yorkers take pride in the Metropolitan Museum's Rembrandts. From this mushroom and its relatives come stunning colors used for dyeing wool and other materials.

Also known as: *Dermocybe semisanguinea.*

Where to find: Growing in small scattered groups, on ground or moss in young, moist conifer woods. Reported from Southeast, Southcentral, Interior, and Arctic Alaska.

When to find: Late summer and fall.

Look carefully for all these features:

- Small to medium-size mushroom.
- Cap yellowish to yellow-brown.
- Gills bright deep red, aging rusty brown.
- Stalk yellowish and hairy.
- Veil on young mushroom yellowish and cobweblike, extending from cap edge to stalk, covering young gills.
- Sporeprint rusty brown; spores borne on gills under cap.

Look-alikes: This is the only mushroom I know of with yellow stalk, yellow cap, and red gills. It looks somewhat like other smallish, thin, closely related *Cortinarius* sometimes placed in a separate genus, *Dermocybe.*

Food use: Do not experiment! Edibility unknown, possibly dangerous.

Warning: This mushroom is related to some mushrooms found to contain dangerous toxins. Until recently, most popular guides listed *Cortinarius* as a safe-to-eat genus. Now some species are known to cause death after a latency period of as long as two weeks or more. The *Cortinarius* are the largest genus, with over 1,000 species, most of them untested, edibility unknown. It's wise to consider all of them suspect.

Golden Pholiota
Phaeolepiota aurea (Family: Cortinariaceae)

Everyone comes to Alaska with a wish list. For my brother, up for his first visit, the list read, "See bald eagles, northern lights, a glacier, and a volcano." This could all be done within sight of our home on Kachemak Bay. For 'shroomers coming to Alaska in autumn, the list includes the notation, "Discover Alaskan Gold!"

During a brief spell—as the days grow short and cool before winter closes in—any roadside or lawn with alders or buried alder debris might be aglow with these large, lustrous mushrooms, heavily powdered with gold dust, as irresistible to rub as Aladdin's lamp.

Also known as: *Pholiota aurea, Togaria aurea,* Alaskan Gold.
Where to find: Usually growing in scattered clumps or groups, on rich soil or buried woody debris, under or near hardwood trees, usually alder. Reported from Southeast and Southcentral Alaska.
When to find: Fall.
Look carefully for all these features:
- Medium-size to large mushroom.

- Cap and stalk gold-colored, dry, powdery or grainy to velvety, becoming smooth if rained on or rubbed; cap is often fringed.
- Flesh pale yellow, thick.
- Gills light yellow to cap-colored.
- Ring on stalk is large, long-lasting, cap-colored.
- Sporeprint yellow-brown; spores borne on gills under cap.

Look-alikes: Somewhat resembles sheathed and grainy-coated *Cystoderma* species.

Food use: Edible for some, poisonous for others. Some members of the Alaska Mycological Society eat the Golden Pholiota regularly without incident, but reports persist of gastrointestinal distress caused by this golden beauty.

Caution: The Golden Pholiota causes severe allergic reactions in some people. Another risk of eating this fall mushroom comes from the fact that frost-damaged caps can contain toxins from decomposing tissue. If you're determined to sample this mushroom, collect only young fresh caps before the first fall freeze. Cook them thoroughly. Don't overindulge, and don't serve them to dinner guests.

Gypsy Mushroom
Rozites caperata (Family: Cortinariaceae)

Gypsy Mushrooms are likely to appear at the same time and on the same paths as human travelers. In Alaska this often occurs during cool autumn days along game trails through moist old-growth spruce or on plush sphagnum moss mounds near berry patches.

Marcee Gray, our neighbor, appreciates mushrooms for their invigorating drawing power: they get her out of the house and into the woods. Gypsy Mushrooms are especially attractive because they're easy to find and identify from their distinctive silky glow on the floor of dark, shadowy old spruce forests.

Also known as: *Pholiota caperata*, Granny's Nightcap, *zigeuner* ("gypsy" in German).

Where to find: Growing in small groups or scattered, on humus or sphagnum moss in moist open places, along trails and ridges among old-growth spruce, in some birch forests, on tundra near dwarf birch. Reported from Southeast, Southcentral, Interior, and Arctic Alaska.

When to find: Late summer and fall.

Look carefully for all these features:
- Medium-size mushroom.
- Cap yellow-orange-brownish, covered with a subtle whitish coating when young; dry, wrinkled near edge.
- Flesh white, thick, firm; not changing color when cut.
- Gills pale tan to brown, edges often lighter, close together, attached to stalk; covered by a sturdy white to pale yellow membranous veil when young.
- Stalk whitish, solid, circled midway by a persistent membranous white to pale yellow ring.
- Sporeprint rusty brown; spores borne on gills under cap.

Look-alikes: Various other brownish-spored mushrooms like *Cortinarius*, *Pholiota*, and *Agrocybe* species, many of which are poisonous or unknown.

Food use: Edible, with caution. The Gypsy is widely considered a safe, delicious edible for most people. It is sold in European and Japanese markets, but it has received little honor in its North American range. It can be dried or pickled for storage. Discard tough stalks before cooking.

Caution: The Gypsy isn't a foolproof edible for beginning or casual collectors because it resembles and is related to some dangerous look-alikes. Be sure to compare your collections with descriptions in current regional field guides. Seek expert assistance to identify any Gypsy Mushrooms you collect for the table until you can unerringly recognize key features (this may take two or more seasons). Occasionally a Gypsy will turn up that lacks two distinctive features: the wrinkles on the cap and the cap's faint whitish sheen. Discard any irregular-looking ones as possible impostors.

Poison Pax
POISONOUS
Paxillus involutus (Family: Paxillaceae)

A pparently no fatalities due to mushroom poisoning have been reported in Alaska. This probably reflects the small number of people who harvest wild mushrooms here and the scarcity of deadly *Amanita* species, which are mistaken for edible look-alikes frequently in other places.

But this does not signify any lack of fungal thugs among us

in Alaska. Less than a mile from where I sit lives Poison Pax, *Paxillus involutus*, responsible for the only recorded fatal mushroom-poisoning of a professional mycologist. Far from being a recluse, Poison Pax is all over town: on the post office lawn, behind the library, and next to the sandbox in the children's park!

Should you glimpse the polished brown-gold caps of Poison Pax on your rounds, take time to stoop and study their intricately symmetrical tidiness. Note their warm, oddly reptilian hues. Then snap to and remember to guard your young children and puppies and make sure you always consult current mushroom guides!

Where to find: Growing grouped or scattered, on ground or rotting wood near trees, often spruce or birch. Reported from Southeast, Southcentral, Interior, and Arctic Alaska.
When to find: Summer and fall.
Look carefully for all these features:
- Medium-size to large mushroom.
- Cap brown, often tinted olive; slightly fibrous; shiny when dry, sticky when wet.
- Cap edge in-rolled and lined or furrowed.
- Flesh yellow, firm.
- Gills pale yellow, aging brown; may be forked or pore-like near stalk; running down stalk; separate readily from cap.
- Stalk yellow-brown, solid, dry, smooth, usually central.
- All parts might stain brown with age or when rubbed.
- Sporeprint yellow-brown; spores borne on gills under cap.

Look-alikes: Similarly colored *Lactarius* and *Russula* species.
Warning: All but recently written mushroom guides list Poison Pax as edible, edible when cooked, or edibility unknown. This mushroom is now known to cause life-threatening gastric disturbances when not cooked sufficiently and fatal allergic reactions in sensitized individuals even when it is cooked thoroughly. Apparently a person can eat these mushrooms for years before the body begins producing antibodies that can destroy red blood cells and cause kidney failure.

False Chanterelle
Hygrophoropsis aurantiaca
(Family: Paxillaceae)

The False Chanterelle is the most commonly seen fungus in Homer during summer and fall. We've recommended in jest that it be designated our official municipal mushroom for its diligence in gilding almost every rotting stump in town with its pumpkin-hued autumn ambience.

This mushroom's common name refers to its resemblance to the chanterelle family of mushrooms, to which it once belonged. It has suffered a perpetual identity crisis at the hands of taxonomists, who have booted it out of two different fungal families this century. And it might well not survive its latest placement with the *Paxillaceae*.

Also known as: *Cantharellus aurantiacus, Clitocybe aurantiaca.*
Where to find: Growing scattered or in groups, on humus or rotting wood near conifers in forests, parks, and lawns. Reported from Southeast, Southcentral, and Interior Alaska.

When to find: Summer and fall.
Look carefully for all these features:
- Small to medium-size mushroom.
- Cap orange, yellow, brown, or cream; dry, suedelike; shallowly depressed in center, with in-rolled margin.
- Flesh pale orange or cap color, thin.
- Gills orange, repeatedly forked, sometimes blunt-edged, running down stalk.
- Stalk orange or cap color, dry.
- Sporeprint white; spores borne on gills under cap.

Look-alikes: *Cantharellus cibarius, Omphalotus illudens*.

Food use: Not recommended. Edibility unknown. There's no consensus in the literature about this mushroom's edibility. Though some people indulge without consequence, persistent reports of gastric illness, indigestion, and hallucinations mar its record. Some suggest that reported poisonings by the False Chanterelle are actually by a misidentified look-alike, *Omphalotus illudens*. I don't pioneer for new edible species and don't advise others to be so bold. Though a pleasantly familiar presence, the False Chanterelle remains a stranger among us, this mushroom that won't sit still for the taxonomists.

Boletes

Boletes are fleshy mushrooms with a cap and central stalk like most gilled mushrooms. Their spores are borne in tubes and released from pores beneath the cap. The spongy tube layer is usually easy to separate from the cap's flesh. Most boletes grow on the ground in association with certain trees. This group contains some very tasty edibles, a few poisonous species, and a number of unknowns. Many sources advise people to avoid eating boletes with pink to red pores, especially those with flesh or pores that turn blue when handled.

Cap is soft and fleshy, like most gilled mushrooms.

Pore surface is a soft and spongy tube layer underneath the cap, a feature that distinguishes boletes from gilled mushrooms.

Stalk base

King Bolete
Boletus edulis (Family: Boletaceae)

I don't know where to begin singing praises of this mushroom. King Bolete is probably the best-loved, most widely eaten, and among the most digestible of wild mushrooms. It is favored by meat eaters and vegetarians alike for its juicy, tenderly chewy texture and sweet, mildly meaty taste. ("Close your eyes and tell which bite is more exquisite. That one? *Mais oui!* Madame will not have the filet mignon this evening!")

The King Bolete can be tricky for beginners to identify. Nineteenth-century mycologist Charles McIlvaine said it well: "Some species of fungi appear to have that prize of fairyland—the Wishing Cap—and by its magic power be able to take on any form they please. *Boletus edulis* is one of them."

The King Bolete not only wears crowns of many colors, but can also vary in size, shape, and habitat. But don't despair. With help from local experts and regional field guides, you'll soon be spotting them at the edge of the forest along the winding Alaskan highways and byways.

Also known as: In Europe, though highly esteemed, it has tags less pretentious than King—in Britain, Penny Bun; in France, *cèpes*; in Germany, *steinpilz* ("stone mushroom"); in Italy, *porcini* ("little pigs").
Where to find: Growing alone, scattered, or in small groups, on ground near conifers or hardwoods. Reported from Southeast, Southcentral, Interior, and Arctic Alaska.
When to find: Late summer and fall.
Look carefully for all these features:
- Large robust mushroom.
- Cap pale yellow-brown to dark red-brown.
- Cap underside (pore surface) spongy, white when young, aging greenish yellow; pores round, tiny.
- Flesh white, not changing color when handled or cut.
- Stalk thick, white or brown, straight or enlarged toward base.

- Upper stalk covered with fine whitish netlike pattern.
- Odor and taste pleasant and mild. Caution: Spit out after tasting a small piece, as raw flesh can cause digestive disturbances.
- Sporeprint olive-brown; spores borne in tubes under cap.

Look-alikes: Other boletes, some poisonous.

Harvesting tips: Collect King Boletes when they are young and fresh. Key features are most easily recognized then, and you'll have a better chance of getting to the Kings before the maggots do. Slice mushroom from base through cap to examine for bugs or spoilage before eating.

Food use: Edible and choice. In ancient Rome, only members of the patrician class were permitted this delicacy. The King rules almost any meal in its presence: sautéed or baked in omelettes and casseroles, simmered in pasta sauces, broiled and served in steaklike slabs. Sliced thin, dried, and stored in jars, Kings keep well for over a year. They provide a rich flavoring agent and a significant protein boost for soups, stews, and stir-fries. Dried Kings are said to be richer in protein than any dried vegetable. Drying intensifies the flavor, and I prefer drying over freezing for this reason.

Caution: King Bolete can cause stomach distress when eaten without cooking thoroughly.

The King has nasty look-alikes. Confirm your identification in several regional guides and run your catch by an experienced collector before sampling a small portion.

Birch Bolete
Leccinum scabrum (Family: Boletaceae)

The showy *Leccinum*s are easily recognized as a clan. Their attractively colored orange, red, or brown caps and solid white stalks decked out in brown or coal black scales are a common sight in summer and fall forests and on tundra, where they are associated with dwarf birch and other birch species.

Though some species are difficult to identify, none of the *Leccinum*s are believed to be dangerously poisonous. It's always a pleasure to find them sparking up the woods each season and to note on my calendar, "They're back!" The Birch Bolete pictured here is one of the more understated of the *Leccinum* jewels, blending without a fuss into its late summer backdrop.

Where to find: Growing alone or in small groups, on ground near birch in forests, bogs, parks, and lawns. Reported from Southeast, Southcentral, Interior, and Arctic Alaska.
When to find: Late summer and fall.
Look carefully for all these features:
- Medium-size to large mushroom.
- Cap brown.

- Cap underside (pore surface) spongy, white to brown; does not turn blue when rubbed or cut.
- Stalk white, coated with brown or black scales; sometimes turning blue-green at base.
- Flesh white, with little or no color change if broken or rubbed.
- Sporeprint brown; spores borne in tubes under cap.

Look-alikes: Other brownish *Leccinum*s.

Food use: Edible. Rated good to choice, with caution.

Caution: Collect and cook when fresh and firm, avoiding specimens with maggots. As with all mushrooms, sample sparingly when eating for the first time. Some individuals have unpleasant allergic reactions to *Leccinum*.

Orange Birch Bolete
Leccinum testaceoscabrum
(Family: Boletaceae)

From the time the Orange Birch Bolete emerges as a young mushroom, its white stalk is densely and handsomely coated with black scales. You could almost imagine it had just emerged from a coal seam in one of the cliffs near its home.

Also known as: *Leccinum aurantiacum*.

Where to find: Growing alone or in small groups, on ground near birch, within or at edges of birch forests and mixed birch/conifer woods. Reported from Southeast, Southcentral, and Interior Alaska.

When to find: Midsummer through fall freeze.

Look carefully for all these features:
- Medium-size to large mushroom.
- Cap bright orange; dry; thin tissue draped over young cap edge.
- Cap underside (pore surface) spongy, olive-grayish.
- Flesh white, firm, turns dark when cut or cooked.
- Stalk white, densely coated with black scales; might turn bluish near base.

- Sporeprint brown; spores borne in tubes under cap.

Look-alikes: Other similar orange-capped, black scale-stalked edible *Leccinum*s.

Food use: Edible, with caution. This mushroom is considered fair to good by consumers in our community. We ate it for a while, but left it when the sensational King Bolete turned up on our rounds. Its habit of turning dark when cut or cooked is unnerving to many diners, though its color and texture can be appealing when served in oriental stir-fries or soups. The Orange Birch Bolete keeps well when it is dried.

Caution: Some people report unpleasant reactions when eating these mushrooms. Always cook thoroughly and sample a small portion when eating these or any mushroom for the first time. Individual allergic reactions can occur with almost any edible mushroom.

Shelf Mushrooms

Shelf mushrooms, also called polypores, can be fleshy and soft or tough to woody. Their spores are borne in tubes and released from pores on the underside of the mushroom. Shelf mushrooms grow on living or dead wood. Some have medicinal value or a history of folk uses. A few shelf mushrooms are popular edibles, but the woody texture and bitter taste of most species render them inedible. Most people feel little appreciation for parasitic shelf mushrooms that may be growing on their favorite trees.

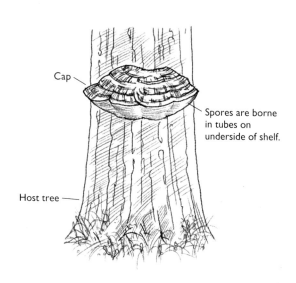

Cap

Spores are borne in tubes on underside of shelf.

Host tree

Sulfur Shelf
Laetiporus sulphureus
(Family: Polyporaceae)

There's something unsettlingly Halloweenish about the Sulfur Shelf's bright orange and yellow flaming presence, with goblinlike knobby fingers creeping out of a dark hollow in a dying tree.

A powerful, persistent parasite, Sulfur Shelf requires little moisture and can fruit when and where most large fleshy

fungi fail. Sulfur Shelf produces mushrooms year after year as long as its host tree provides nutrients and remains intact. This is a fine arrangement for Sulfur Shelf and its 'shroomer fans, and for critters that inhabit hollow trees, but not for the host tree and for the humans who may value the tree for shade and beauty or construction materials.

Sulfur Shelf's appetite for wood results in heart rot for its victim, and the hollowed tree is eventually toppled by wind. Boats made from infected wood soon become unseaworthy.

Also known as: *Polyporus sulphureus*, Chicken-of-the-Woods.

Where to find: Often growing in overlapping clusters on stumps, logs, and trunks of spruce, hemlock, and cottonwood trees. Reported from Southeast and Southcentral Alaska.

When to find: Late summer and fall.

Look carefully for all these features:
- Medium-size to large shelflike caps.
- Cap upper surface brilliant orange and yellow; smooth or suedelike; wrinkled and wavy toward cap edge (usually yellow).
- Flesh yellowish white, tender and juicy, oozes yellowish droplets, ages tough and brittle.
- Undersurface (pore surface) sulfur yellow when fresh.
- Odor musky.
- Taste tart.
- Sporeprint white; spores borne in tubes under shelf.

Harvesting tip: Use sharp knife to slice caps off near wood. Young caps require little cleaning, because insects aren't usually attracted to them.

Food use: Edibility questionable. Long considered safe and choice, this popular, widely eaten fungus is now known to cause digestive disturbances. Some instances of poisoning, causing such symptoms as swollen lips, appear to be allergic reactions. Others might be related to the type of tree the fungus was growing on (eucalyptus, hemlock, and various conifers are suspects). Contamination by air or soil could be a factor, as could drinking alcoholic beverages with a meal of these mushrooms. Still fans of Sulfur Shelf continue to feast on their fickle favorite at home and in fine restaurants. The chickenlike texture and taste more than adequately replace chicken in many gourmet poultry dishes and salads. In stews, barbecues, and stir-fries, the Sulfur Shelf adopts the chewy meatiness of beef.

Caution: If you can't walk away from this question mark, do take a few precautions. Collect only fresh, new, tender growth at the outer edge of young caps. Clean and cook thoroughly (never eat them raw). Eat a very small portion when first sampling Sulfur Shelf, and eat prudent-sized portions thereafter.

Birch Conk
Piptoporus betulinus (Family: Polyporaceae)

Lovers of birches might shudder to see this enemy climbing tree trunks, for the Birch Conk produces a crumbly rot that slowly destroys its host. Yet people raised on Yankee homesteads in the early 1900s found this lowly conk to be almost indispensable. It was used as an anesthetic and antibiotic, as well as for making fire starters and razor strops.

The Birch Conk's service to humanity actually goes back much further than our grandparents' farms. *National Geographic* magazine reported on the remains of a Bronze Age man found in a glacier high in the Alps. Found with him was his trusty Birch Conk, perhaps packed as an emergency kit to provide starters for fires or an antibiotic for dressing wounds.

Also known as: Birch Polypore, Razor Strop Fungus, *Polyporus betulinus*.
Where to find: Growing alone or in groups, on living or dead birch trees. Reported from Southeast, Southcentral, Interior, and Arctic Alaska.
When to find: Fruits in summer and fall, but remains intact a long time and can be seen on trees year-round.
Look carefully for all these features:
- Small to large fungus.
- Shape semicircular or hooflike.

- Cap upper surface whitish tan, gray, or brown; smooth, with in-curving thick margin.
- Undersurface (pore surface) white, aging tannish.
- Flesh thick, white, firm, moist, somewhat rubbery, aging dry and corky.
- Stalk stubby; often absent.
- Sporeprint white; spores borne in tubes under shelf.

Food use: Not recommended. This conk remains out of sight, out of reach, and out of mind for most potential tasters and isn't eaten widely enough to assess its edibility. A few reports dub it edible when young, tending toward bitter and tough.

Artist's Conk

Ganoderma applanatum
(Family: Polyporaceae)

The Artist's Conk proves that one primate's palette pleases another primate's palate! Dian Fossey in *Gorillas in the Mist* describes the humanly inedible Artist's Conk as a rare treat among the mountain gorillas of Rwanda. Only the oldest and strongest manage to pry it off trees, and then they must protect their prize from more dominant animals.

Fossey tells of an outing with two young orphan gorillas, Pucker and Coco, which turned into an adventure that 'shroomers of any species can relate to. The two youngsters suddenly became agitated and raced toward some distant *Hagenia* trees and stopped beneath one. She writes: "They peered up at the tree like children looking up the chimney on Christmas Eve." Then they scrambled up the tree, grunted at each other, and hung on by arms and legs while they nibbled away at a huge Artist's Conk until scarcely a smear remained.

The common name of this mushroom refers to use of the white pore surface as an etching pad. When this under-

surface is scratched with needle, knife, or fingernail, the scratches immediately turn permanently brown. I've seen some lovely scenes, maps, designs, and all manner of messages etched on this conk.

In Chinese medicine, various *Ganoderma* species, especially *Ganoderma lucidum*, are valued for properties that are described as anti-tumor and immunity-building.

Also known as: *Fomes applanatus.*

Where to find: Growing alone or severally, on living or dead conifers or hardwoods. Reported from Southeast, Southcentral, and Interior Alaska.

When to find: Year-round; they're perennial. These mushrooms add growth layers rather than producing new shelves each year.

Look carefully for all these features:
- Medium-size to large, hard, shelf-shaped woody fungus.
- Upper surface has dull tan, gray, or brown furrowed crust.
- Undersurface (pore surface) white when young. Fresh underside turns brown and remains so wherever it is scratched or handled, and ages reddish brown.
- Flesh woody, showing brown annual layers if broken in half.
- Stalkless.
- Sporeprint brown; spores borne in tubes under shelf. Spores numerous (billions released daily). They can be seen as dense powder on nearby caps and on host tree's trunk.

Look-alikes: *Fomitopsis pinicola* and other large flat woody Conks.

Food use: Inedible (but gorillas beg to differ).

Caution: Artist's Conks around the house might release enough spores to endanger health. Some folks spray them with clear fixatives, available from art or drafting suppliers, to prevent spread of spores.

Teeth Mushrooms

Teeth mushrooms can be found on the ground, with caps and stalks that range from tough and leathery to fleshy and brittle. The spores are borne on spines (teeth) on the underside of the caps. Other kinds of teeth mushrooms can be found on dead or living trees, where they appear as delicately beautiful masses of bright white or cream-colored growth. In these tree dwellers, the spores are borne on iciclelike projections hanging from the fungal "branches" or fleshy bases. Teeth mushrooms grow mostly in northern conifer forests. Some are excellent edibles; others are tough and bitter.

Ground-dwelling type:

Tree-dwelling type:

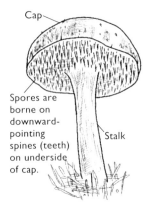

Cap

Spores are borne on downward-pointing spines (teeth) on underside of cap.

Stalk

Spores are borne on hanging projections.

Iciclelike spines or teeth hang from branches or fleshy bases without well-defined cap or stalk.

Hedgehog Mushroom
Hydnum repandum (Family: Hydnaceae)

The Hedgehog is a favorite, much-sought edible mushroom in our neck of the woods. This tasty forest food is easy to know and, once found, is easy to find again each season in the same place.

It's considered gauche at best to inquire more than generally about the location of a neighbor's Pumpkin Pig patch. As the population of harvesters grows in Alaska, people naturally get quieter about their favorite hunting, fishing, berrying, and 'shrooming places. And if you think it's unfriendly to hog the resources and hedge about their whereabouts, just wait till you meet the morel monopoly in May!

Also known as: *Dentinum repandum*, Pumpkin Pig.
Where to find: Growing alone, scattered, or in groups, on ground near spruce or other conifers. Reported from Southeast and Southcentral Alaska.
When to find: Fall.
Look carefully for all these features:
- Small to medium-size mushroom.
- Cap commonly cream to pale orange, aging or bruising dark orangish brown; smooth and dry.
- Underside has white or pale cap-colored short teeth or spines instead of gills or pores.

- Flesh white and brittle.
- Stalk cream-colored or cap-colored; smooth and dry.
- Odor fruity.
- Sporeprint white; spores borne on spines.

Look-alikes: Some similar-looking teeth mushrooms are bitter tasting, though none are known to be poisonous. The Hedgehog sometimes has a smaller sidekick, *Hydnum umbilicatum* (the "belly button" *Hydnum*), appearing like a Hedge-Piglet alongside its close relative. The Belly Button, with a depressed cap center suggesting a navel, has a white sporeprint and is edible, just like the Hedgehog.

Harvesting tip: Leave the stalk base in the ground to ensure future crops. Mycelium can be damaged by greedy 'shroomers rooting around for every last bite in a patch!

Food use: Edible; excellent taste and texture. Hedgehogs are a widely collected edible forest fungus throughout their North American and European range. They are an easily identified and safe edible for most people. When collected fresh, examined for rarely found maggots, and cooked soon after in casseroles, sauces, or gravies, Hedgehogs transform the most modest meal into a gourmet dining fête. When sautéed in butter or olive oil and frozen, they retain their flavor and texture for several months. Older Hedgehogs should be boiled, drained, and cooked in milk to minimize a possible bitter taste.

TEETH MUSHROOMS

Comb Hericium
Hericium ramosum (Family: Hydnaceae)

The splendid sight pictured here has lived for years on a dead cottonwood in a narrow stretch of mixed woods by Kachemak Bay. It wouldn't be startling to learn that it's actually some rare sea coral swept up, log and all, by a tidal wave and left high to dry.

The feathery pristine presence of this *Hericium* suggests to some admirers an exotic albino fern or an ice crystal gone haywire—or an orderly maze of fine lacework hastily dropped by fairies when they heard us coming!

Also known as: *Hericium laciniatum.*

Where to find: Growing alone or in small groups, on hardwood logs, stumps, or fallen branches. Reported from Southeast, Southcentral, and Interior Alaska.

When to find: Late summer and fall.

Look carefully for all these features:

- Large, densely branching mass of fungus.
- Colored white throughout the fresh fruiting body, including spines, flesh, branches, and base attachment; aging yellowish to brownish.
- Delicate, slender branches emerge from rootlike base.
- Spines (or teeth) are short, and arranged comblike along sides of the branches and toothbrushlike in tufts at tips.
- Sporeprint white; spores borne on spines.

Look-alikes: Other *Hericium* species, especially *Hericium abietis*, which grows on spruce. No poisonous look-alikes.

Harvesting tip: Collect white, fresh combs and place them alone in a clean basket.

Food use: Edible; a delectable delicacy. This is a good-news fungus: beautiful, safe, delicious, durable, fruiting at the same time on the same log for years! For best taste, texture, and digestive success, select fresh combs, then clean them and cook slowly. The *Hericiums* are keepers. They are successfully cultivated on wood to provide fresh mushrooms, and they make elegant crisp pickles.

Like many fungi, the *Hericiums* are being studied for medical uses. An extract from *Hericium* tissue has shown antitumor activity in cancer research with mice.

Coral Mushrooms

Coral mushrooms are soft-fleshed and attractively colored. Their spores are borne on single or branching surfaces of the clublike or coral-like mushrooms. Corals thrive on the ground or on dead wood, often in northern or mountain coniferous forests. Some are fine edibles; others induce nausea.

Spores are borne on surfaces of upright clubs or coral-like branches.

Strap Coral

Clavariadelphus ligula and *Clavariadelphus sachalinensis* (Family: Clavariaceae)

The first time I saw Strap Coral I thought some strange plant was sending up shoots all over the woods near my house. Only on looking closer did I realize this was a very prolific fungus. I referred to these species as "Tongues" long before I properly identified them. Had such Tongues voices, the woods might ring with a thousandfold harmonies.

On the other hand, perhaps another common name, Clubs, is applicable. Looking at them, you might imagine a forest of baseball bats—or the beginnings of a barbaric army from the myths of yore.

Also known as: Clubs, Tongues.
Where to find: Growing scattered or in groups, on mossy ground in well-drained conifer forests. Reported from Southcentral Alaska.
When to find: Late summer and fall.
Look carefully for all these features:
- Small mushroom.
- Cap and stalk are one; pale yellowish to pale brownish, club-shaped, lightly wrinkled lengthwise.

- Flesh white to yellowish.
- Sporeprint white to pale yellowish for *Clavariadelphus ligula*, darker for *Clavariadelphus sachalinensis*; spores borne on surface.

Look-alikes: Other small unbranched coral mushrooms, particularly *Clavariadelphus pistillaris*, which prefers hard-woods and mixed-woods habitat.

Food use: Not recommended. Some people say all *Clavariadelphus* species are edible, others say they're not. However, no one says they're choice or great, and no one spends much time picking them if anything else is waiting in the woods.

Puffballs

The puffball is a spore case that carries the spore mass—the mushroom's reproductive units. Inside the puffball, the spore mass changes from a firm cheesy substance into the powderlike mature spores that escape from a tiny hole, slit, or tear in the top of the puffball. The spores are dispersed by wind, rain, and animals, some coming to rest in suitable niches for making more puffballs.

The scientific family name of puffballs, *Lycoperdaceae*, translates as "wolf fart." I didn't get it until the day I watched kids in a puffball patch stomping and yelling "Oh,

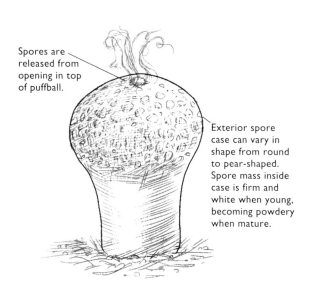

Spores are released from opening in top of puffball.

Exterior spore case can vary in shape from round to pear-shaped. Spore mass inside case is firm and white when young, becoming powdery when mature.

gross!" over and over until the last poof. Thus puffballs and human youngsters meet perennially on the same stomping grounds, to the delight of both children and puffballs. (But read the following cautions, sigh.)

Puffballs have a history of human use in addition to being edible. They were used by Native Americans to stop bleeding, to treat tumors and inflammation, and as a dusting powder for babies. Surgeons in the nineteenth century used the spores and slices of fresh puffballs to dress wounds. Currently, tumor-inhibiting derivatives from some puffball species are being tested in cancer research.

With caution, most people can identify and safely eat the puffballs featured in this section. Observe the following cautions for puffballs:

- Slice open every puffball lengthwise before cooking. The inside of a fresh, edible puffball will be firm yet slightly spongy, white, with homogeneous texture throughout.
- Discard any puffballs that are at all yellowish green, mushy inside, or infested with maggots. These can cause nausea and a single rotten ball can spoil a stew.
- Beware of look-alikes:
  Amanita "eggs" or buttons, when sliced open lengthwise, show an outline of the embryonic cap, gills, and stalk. These might be deadly poisonous.
  Earthballs (*Scleroderma* species) have tough skin and a hard white spore mass that turns hard black before becoming brown-black and powdery at maturity. These mushrooms can cause severe gastrointestinal distress.
  Other similar-appearing puffballs (some edible, some unpalatable, others unknown). *Identify to species any mushrooms you intend to eat.*
- Eat sparingly when first trying puffballs, as you would any wild mushroom new to you. Some people experience digestive disturbances from puffballs.
- At the risk of being considered a killjoy, it's my duty to inform you that the pleasure of kicking puffballs may be hazardous to your health. Inhaling the spores that are released has caused bronchial irritation and allergic reactions. Perhaps this is how the puffball came by its nickname, Devil's Snuffbox.

Tumbleball

Bovista plumbea (Family: Lycoperdaceae)

We often eat fried sliced Tumbleballs. While not a gourmet gem, they're quite good, and we look forward to finding them while mowing the lawn. From mid-summer until fall freeze, we find fresh balls at almost every cutting, especially after rain. They're golf-ball size, with the occasional tennis ball or baseball monster thrown in. Yet they can rarely be seen except by the person mowing, who can look straight down and see their little heads pushing the grass aside.

Tumbleballs' papery metallic-looking spore cases, often still half-filled with spores, can be found year-round, strewn about over bare lawns, fields, and footpaths—especially early spring after snowmelts.

Also known as: True Puffball, Tumbling Puffball.
Where to find: On ground in fields, lawns, and other open grassy places. Reported from Southcentral Alaska.
When to find: Summer through fall.
Look carefully for all these features:
- Small to medium-sized round puffball, often with a dirty patch of fibers at base.

- Skin white and smooth when young, shedding to reveal a papery metallic gray-brown inner skin that eventually falls apart, releasing spores.
- Spore mass firm and white; then becoming yellowish green and mushy; turning dark brown and powdery when mature.
- Spores brown.

Look-alikes: The closely related species *Bovista pila*; other round puffballs; button stages of some gilled mushrooms, including *Amanitas*, which can be deadly poisonous.

Food use: Edible for most people. These puffballs are pleasant eating and safe in moderation. Slice them open and discard any specimens that are soft or discolored from ripening. (See the general cautions for puffballs.)

Gemmed Puffball
Lycoperdon perlatum
(Family: Lycoperdaceae)

Many people cut their 'shrooming teeth on these puffballs. They're nearly everywhere in lawns, parks, ballfields, and woodlands, and they provide a familiar game of romp-and-stomp for children.

When Gemmed Puffballs live up to their name, the outer layer of skin—composed of tiny spines shaped like cones and pyramids—sparkles in the sun. Looking at one can be like gazing for a magic moment at the crown jewels.

Also known as: *Lycoperdon gemmatum*; Devil's Snuffbox.
Where to find: Growing in small groups or clustered on soil or humus in grass and woodlands, and on lawns. Reported from Southcentral Alaska.
When to find: Late summer to fall freeze.
Look carefully for all these features:
- Small, white, pear-shaped puffball.
- Covered with glistening scales, conical and pyramid-shaped, that fall off and leave smooth "pockmarks" with age.

- No true stalk, but a stalklike chambered base.
- Spore mass firm and white; then turning yellowish-green and soft; becoming powdery and olive-brown at maturity, before releasing spores from pore or slit at top of puffball.
- Spores brown.

Look-alikes: Other puffballs; button stages of some gilled mushrooms, including *Amanitas*, which can be deadly poisonous. (See the general cautions for puffballs.)

Food use: Edible for most people. Reports vary from good to bland to bitter. The precaution of selecting only fresh puffs with no trace of yellow inside should prevent bitter taste and gastric gripes. (See the general cautions for puffballs.)

Pear Puffball
Lycoperdon pyriforme
(Family: Lycoperdaceae)

The Pear Puffball is common in our woods and wherever rotten wood may have been shallowly buried in our rush to clear homesteads or plant lawns and gardens. Woody-colored and densely packed on its look-alike habitat, it's discovered more by chance than by hunting.

Where to find: Growing scattered to densely clustered in lawns, fields, and mixed woods, and on stumps, logs, buried wood, and sometimes dead standing trees. Reported from Southeast, Southcentral, Interior, and Arctic Alaska.
When to find: Fall.
Look carefully for all these features:
- Small, smoothish, pale brown, pear-shaped puffball.
- Short, white, stalklike chambered base.

- Visible white strands of mycelium connecting base to surrounding rotten wood.
- Spore mass firm and white; then turning yellowish green and soft; becoming olive-brown and powdery before releasing spores from pore or slit at top of puff-ball.
- Spores brown.

Look-alikes: Other *Lycoperdons*; button stages of some gilled mushrooms, including *Amanitas*, which can be deadly poisonous. (See the general cautions for puffballs.)

Food use: Edible for most people. Usually good when fresh, but discard balls with yellow or other discoloring spore mass. One bad Pear spoils the barrel. (See the general cautions for puffballs.)

Morels and False Morels

Morels and false morels have a thin, spongy cap and stalk. They are colored white or cream to brownish black. Spores are borne in the pits of morel caps and on the surfaces of false morel caps. Morels can be found in spring in diverse places, including burned or unburned forests or fields, logged areas, orchards, gardens, and hillsides. False morels occur from spring to early fall on the ground or on rotten wood in forests, along trails, and in clearings. The "true" morels are among the choicest edible mushrooms; some of the false morels are dangerously poisonous.

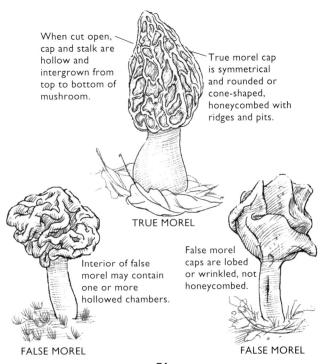

When cut open, cap and stalk are hollow and intergrown from top to bottom of mushroom.

True morel cap is symmetrical and rounded or cone-shaped, honeycombed with ridges and pits.

TRUE MOREL

Interior of false morel may contain one or more hollowed chambers.

False morel caps are lobed or wrinkled, not honeycombed.

FALSE MOREL

FALSE MOREL

Black Morel
Morchella angusticeps and closely related
species (Family: Morchellaceae)

The first spring I lived in Alaska, I found my morel places smack where and when the books said to look: on south-facing slopes in May and June, after the snow melts and the air becomes mild. Since then I've described in detail the habitat and done everything short of drawing a map and putting up road signs—and still there are people who forage successfully for other mushrooms but have yet to taste a morel they picked themselves.

Why they don't find them is simple, but not simple to fix. It isn't so much *where* to find morels as *how* to find morels that stumps would-be morelivores. You must look with eyes grown wise to their camouflage tactics.

The Black Morel does not reflect light. Like a black bear at dusk, this morel is seen only because it's less visible than its surroundings. When you first begin to search, you must strain to see it. I spotted my first morel in Alaska near the trunk of a cottonwood tree because the morel suddenly "moved." That is, my eye caught it before it could blend back into its surroundings.

I realized this was no ordinary 'shroom. No bright reds, yellows, or whites bouncing noisily off the forest floor. Here was a fungus that, like deer, had evolved to merge seamlessly with its surroundings. I looked away and looked back. It was gone! On hands and knees, I bowed my head and stared hard directly at the trunk until the morel's head "moved" again—then *pounce*! Into the basket and slam the lid.

After a few such lessons I learned the humble habits of successful hunters: creep slowly along on all fours, nose to the ground, eyes alert. Forget your pride.

There's hope for those without the predator's knack for stalking morels. Commercial cultivation is being attempted by European and American teams (including a well-known pizza company). Kits are being marketed for growing morels on a small scale, and they work sometimes. And some lucky people have actually "planted" morels by pouring wash water from a few collected morels (spore stew) on their vegetable gardens or compost piles or beneath appropriate trees on their lawns.

The term "black morel" is often used to refer collectively to a group of mushrooms that look much alike and that occur together and appear to hybridize. Some mycologists believe they are all variations of the same species. To foragers it matters only that all of the *Morchella*s are edible and tasty when the usual cautions are observed.

Where to find: Growing scattered, sparsely or abundantly, in open woods and their edges among alder, cottonwood, aspen, and conifers; on south-facing slopes in canyons and on hillsides; in burned-over fields and forests a year or so after fire; in recently logged areas; in compost and gardens, especially if mulched with rotting wood; and in some surprising places. Reported from Southcentral and Interior Alaska.

When to find: May and June, after snow melts and night temperatures become mild. In Southcentral Alaska, often coincides with strawberry flowering.

Look carefully for all these features:
- Small to medium-size mushroom.
- Cap with gray-black ridges framing amber-colored pits, conical to round-shaped, hollow.
- Stalk white or buff, furrowed, grainy textured, hollow.
- Cap and stalk intergrown and hollow inside from top of cap to near base of stalk.
- Sporeprint inapplicable to this group. Not useful for field identification.

Look-alikes: Other *Morchella* mushrooms; and *Verpa Gyromitra*, *Helvella* mushrooms.

Harvesting tip: Slice each morel open from head to toe as you collect it, and use a small brush to clear out any insects or other debris. This will help keep your collection clean, fresh, and appetizing.

Food use: Edible and choice, with caution. In his *Field Guide to Western Mushrooms*, mycologist Alexander Smith writes of our prized Black Morel, "Edibility: Questionable"! A Northwest mycologist informed him that these mushrooms caused more poisonings in his area than any other. My

review of the literature suggests that while a sizable number of people suffer gastrointestinal reactions from eating Black Morels, most people can eat them safely if they observe the cautions described here.

Caution: Be certain you've correctly identified your collection as true morels (*Morcella* species). A true morel cap is symmetrical and rounded or cone-shaped, with a honeycomb or netlike pattern of pits and ridges. The cap and stalk are hollow and intergrown from top to bottom of mushroom. False morel caps are lobed or wrinkled. Their stalk interior may contain one or more small hollowed chambers.

Be certain they're fresh, and clean each one inside and out before cooking. As with many edible mushrooms, some people suffer mild to severe reactions when consuming the choice morel species. Always cook morels thoroughly; never eat them raw. Eat very small portions to begin with; never overindulge. Some sources suggest not drinking alcohol when eating morels or avoiding using alcohol when cooking them. Other sources advocate parboiling morels and discarding the water before using them in recipes.

Brain Mushroom
POISONOUS
Gyromitra esculenta (Family: Helvellaceae)

Aptly named, the Brain Mushroom in its maturity is deeply convoluted, though it lacks the brain's symmetry. Whenever I come upon these grotesque, cerebral parodies in the woods or by a trail, I sense something is out of whack. This may be a healthy reaction. The species name, *esculenta*, translates as "edible," but the Brain is deadly poisonous when eaten raw (and sometimes when cooked), as are most other false morels.

False morels contain the toxin gyromitrin and its byproduct MMH (monomethylhydrazine), which is used as a rocket fuel. Though cooking sometimes reduces the toxins to a safe level, people have become ill simply from inhaling the vapors while simmering these mushrooms.

The difference between a safe and a deadly serving seems to depend on the individual diner's resistance, the concentration of poisons in the mushrooms, and the amount eaten. Death can occur from liver damage or heart failure. Recent animal studies provide evidence of significant carcinogenic properties in gyromitrin as well.

Though related to the true morels, and appearing in the same season and sometimes alongside them, the Brain Mushroom is different enough that collectors familiar with both false and true morel species will not be confused.

Also known as: Lorchel, False Morel.
Where to find: On rotting wood or ground in coniferous or hardwood forest. Reported from Southeast, Southcentral, and Interior Alaska.
When to find: Spring and early summer.
Look carefully for all these features:
- Medium-size mushroom.
- Cap brownish; slightly wrinkled or folding when young, deeply wrinkled in maturity; brainlike.
- Flesh thin.
- Stalk white to light brownish; smooth to lightly wrinkled; usually partly hollow; base of stalk often enlarged.
- Sporeprint not useful for field identification.

Look-alikes: Other *Gyromitra* species, *Helvella* species, and slightly resembling *Morchella* species.
Food use: Deadly poisonous when raw. Not recommended even when thoroughly cooked or dried.
Warning: Eating this mushroom can cause severe illness or death. Breathing fumes from cooking can be injurious.

Hooded False Morel
POISONOUS
Gyromitra infula (Family: Helvellaceae)

The Hooded False Morel has a quaintly elfin aspect and a symmetry its brainy relatives lack. When seen from its own level in the rich green of a miniature mossy glade, by a 'shroomer rolling about on the forest floor, it gives the haunting impression that there really might be leprechauns about. It is typically formed with two or three lobes on a short stalk, each lobe shaped like a monk's cowl. With size and age, the symmetry declines.

Also known as: *Helvella infula*.

Where to find: On decaying wood or ground in coniferous or hardwood forest. Reported from Southeast, Southcentral, Interior, and Arctic Alaska.

When to find: Summer until early fall.

Look carefully for all these features:

- Small to medium-size mushroom.
- Cap brownish, with two or three lobes that are often irregular but tend to be hood- or saddle-shaped; smooth to somewhat wrinkled, but not brainlike.
- Flesh thin, cap-colored, lighter inside.
- Stalk pale to brownish, hollow, smooth to lightly grooved.
- Sporeprint not useful for field identification.

Look-alikes: Other *Gyromitra* and *Helvella* species, particularly *Gyromitra ambigua*. Also slightly resembles *Morchella* species.

Food use: Deadly poisonous when raw. Not recommended even when thoroughly cooked or dried.

Warning: Eating this mushroom can cause severe illness or death. Breathing fumes from cooking can be injurious.

Eating Mushrooms, Alaska-Style

After you've gathered your mushrooms, it's time to put them to use in the kitchen. Here's an informal look at how some Alaskans like to prepare and eat their wild mushrooms.

Preparing Wild Mushrooms for Use

Mushrooms collected in the wild look a lot less clean and tidy on the cutting board than they did in the heat of the chase. First, make sure the 'shrooms you're about to prepare are all the same species and that the species is indeed the edible one that you had in mind.

Next, carefully remove mud, sand, spruce needles, strands of moss, and cowering insects. Most debris is more easily removed dry than wet. You can buy special brushes for this purpose, but we improvise. (A small paintbrush or a sturdy feather works well.)

Finally, cut off and discard any portions of mushrooms that are discolored, infested with maggots, or otherwise suspect. Maggots are fond of many of our edible favorites, particularly the stalk and portions of the cap. While one or two of the little wrigglers doesn't hurt the flavor, they're not appealing, and large numbers of them cause foul odors and major decomposition of the mushroom.

Refrigerating. Depending on the species, mushrooms will keep from one to three days in the refrigerator if they are stored in a paper bag or waxed paper. Do not store mushrooms in plastic bags, because they need some air circulation. I advise cooking all mushrooms before refrigerating to slow decomposition. Inky Caps and Shaggy Manes must be cooked immediately after collecting, unless you're trying to make ink the old-fashioned way.

Freezing. Don't freeze uncooked mushrooms. Instead, follow the recipe later in this chapter for sautéed mushrooms. Let them cool, then package in small freezer bags, being careful to squeeze out all the air before sealing. Label with species and date of freezing on a band of freezer tape wrapped all the way around the bag so it will stay on. Use mushrooms within six months of freezing them. Frozen mushrooms are usually best added to recipes without thawing first. Species we freeze are Hedgehogs, Orange Delicious, and Meadow Mushrooms.

Drying. Some mushrooms are excellent dried, even more flavorful than when they are fresh. Species we especially like dried are morels and the King Bolete. They keep very well.

Prepare the mushrooms as described above. Then slice them thinly, preferably 1/8-inch-thick or less.

Few of us own commercially made food dryers, so we devise our own drying methods. Mushroom slices can be placed on racks or screens (with the exception of stainless steel, metal is not an acceptable surface for drying). Some people dry mushrooms in the oven at the lowest setting, with the door propped slightly open. We use plastic window screens, placed in the house where air circulation is good but where they are out of the sun. Later in the season we hang the screens high near the wood stove. It's important that mushrooms dry fairly promptly, without overheating or becoming damp or dirty.

Mushroom slices should be dried until brittle (or in some cases slightly leathery), shrunken, and virtually weightless. They can then be stored in tightly sealed glass jars, awaiting the cook's desire.

Dried mushrooms are reconstituted by soaking briefly in a minimal amount of water. If they are to be simmered in soup or sauce, reconstitution can be skipped.

Screens and racks must be carefully cleaned, as mushrooms leave a film on them that may mold before the next season.

Recipes

Sautéed Mushrooms
Harriette Parker

This is *the* basic mushroom recipe, and it's the first step in many other mushroom recipes. Vary it to suit your purposes and your chosen mushrooms.

$\frac{1}{2}$ pound clean, fresh mushrooms, wild or domestic
1 to 2 tablespoons butter (or half oil, half butter)
1 garlic clove, crushed (optional)

Very small mushrooms can be cooked whole; otherwise, cut mushrooms in $\frac{1}{8}$-inch-thick slices. Melt butter in a 10-inch frying pan. Add crushed garlic if desired. Add mushrooms. Sauté mushrooms slowly, until tender (10 minutes or more, depending on mushroom species). Some species give off a good deal of liquid, which can be a flavorful addition to sauces, but for other purposes you may wish to simmer until fluids evaporate and mushrooms begin to brown and crisp a bit.

Makes about 2 cups.

Mushroom and Noodle Soup
Neil McArthur

$\frac{1}{2}$ pound clean, fresh mushrooms (your choice)
3 green onions
1 large garlic clove
2 tablespoons butter
4 cups chicken or beef broth (or two 10$\frac{1}{2}$-ounce cans plus one can water)
4 ounces noodles (we use fettuccine or spaghetti, broken into short lengths)
Salt and pepper to taste
Dry sherry (optional)
Parmesan cheese

Chop about a quarter of the mushrooms coarsely; chop the rest finely. Cut green onions into $\frac{1}{2}$-inch-long pieces. Mince garlic. In a medium frying pan, melt butter and sauté mushrooms, onions, and garlic until just cooked. In another pot, bring broth to a boil, then add sautéed items. Bring back to a boil, then add noodles and seasonings. Simmer 10 to 15 minutes, until noodles are cooked. If desired, add a dash of dry sherry toward end of simmering. Sprinkle top of each serving with grated Parmesan cheese.

   Makes about 6 servings.

An Outdoorsman's Seasonal Specials

Bumpo Bremicker

   Like many Alaskans, I like to hunt and fish, and that includes hunting for mushrooms. Here are a couple of my favorite meals.

Morels, King Salmon, and Nettles. In May, after months of eating nothing but fish from the freezer, I get excited about eating fresh king salmon. Fortunately, that's also the time that I can find delicious morels and my favorite wild vegetable: stinging nettles. Such good food needs little to enhance it. Simply sauté sliced morels in a little butter in a black iron skillet along with king salmon steaks. Steam the young nettles (picked with gloves, or carefully with bare fingertips) in a little water until tender. Eat this meal alongside a beautiful stream or while watching the sun set over mountains and ocean as the perfect way to end a day in paradise.

Hedgehogs and Deer. In the fall I hunt for Sitka black-tailed deer. Where I hunt, there are large mossy areas with many Hedgehogs, one of my favorite mushrooms. The mushrooms are well named—not only for the little spines that grow on them but also because they are large and meaty, the hogs of the fungus world. When I return to camp with a deer and some nice Hedgehogs, I break out my cast-iron skillet, get it nice and hot over the campfire, throw in a small piece of deer fat, and wait till the skillet is shiny. I cut my Hedgehogs in chunks, cooking them in the skillet until the water that comes out of them cooks off and they are tender. I set the Hedgehogs aside, throw in

another chunk of fat, and add thin slices of deer backstrap, frying them on each side. When they're done, I pile them on a plate smothered in Hedgehogs. It's time to dig in! It can't get any better than that after a long day in the woods.

Pan-fried Puffballs
Harriette Parker

> Several small clean, fresh puffballs (or one large one)
> 1 egg, beaten
> ½ cup flour or cornmeal
> 1 to 2 tablespoons butter
> Garlic, salt, Parmesan cheese, or other seasonings (optional)

Slice puffballs about ¼-inch thick. Discard any puffballs that are not firm and white inside. Remove skin from edges of the slices. Dip slices in beaten egg and coat lightly with flour or cornmeal. Fry in butter in an open pan over low heat until slices begin to brown and become slightly crisp at the edges. Season as desired.

Serves 2 to 3 as an appetizer.

Spicy Tofu and Hedgehog Stir-fry
Neil McArthur

This is a simple Szechuan-inspired dish prepared from ingredients we usually have around. We make a lot of substitutions, but like to include one protein source, one onion relative, one yellow or white vegetable, and one leafy green vegetable. The overall procedure is to assemble all the ingredients, chop them up as necessary, then cook them in a wok (beginning with the ingredient that takes the longest to cook and ending when everything is cooked just enough).

> ½ cup cooking oil (canola or sesame)
> 3 to 4 medium-size carrots, sliced in ⅛-inch-thick rounds
> 1 leek, sliced crosswise in ¼-inch thick pieces
> 3 to 4 Hedgehogs (preferred because of firm texture and nice color, but others can be used), cut into ½-inch cubes
> 3 tablespoons grated fresh ginger (freeze fresh ginger, then grate)
> 3 or more garlic cloves, minced

1 1/2 teaspoons honey
1/4 teaspoon cayenne
4 to 6 ounces soy sauce
8 to 12 ounces firm (Chinese) tofu, cut in 1/2-inch cubes,
   rinsed, and drained
10 to 15 kale fronds, young enough to be fairly tender;
   remove center ribs and cut into 1-inch pieces (or use half
   of a small savoy cabbage, chopped into 1-inch squares)
1 teaspoon or so cornstarch for thickener, if desired

Heat oil in a wok over high heat until oil just begins to smoke. Cook carrots first, stirring to prevent burning. Add leeks; continue to stir constantly. Follow next with Hedgehog pieces, ginger, garlic, honey, cayenne, and a little soy sauce; keep stirring. Add tofu cubes and more soy sauce. After the contents of the wok regain their heat, add the kale a bit at a time until the last of it is just cooked. To thicken, stir in cornstarch dissolved in a little soy sauce, turn down heat, and cook until mixture has thickened. Serve over rice.

Makes 6 servings.

Moose and Mushroom Stew
Daisy Lee Bitter and Conrad Bitter

2 onions, chopped
1 to 2 pounds moose steak, cut into cubes
1/2 to 1 pound clean, fresh mushrooms
Garden vegetables, chopped
Seasonings to taste

Brown the moose chunks in some oil with onions. Add browned chopped mushrooms (Orange Delicious or King Bolete are our favorites). Add chopped garden vegetables, such as carrots, celery, and potatoes. Make a gravy seasoned with salt, pepper, basil, bay leaf, and a hint of garlic; add it to the stew. Simmer until vegetables are tender.

Makes 4 servings.

Reading and Resources

Note: In determining the edibility of a mushroom, use the most current guides available. The edibility status of any mushroom might change based on new research.

Ammirati, Joseph F., J. A. Traquair, and P. A. Horgen. *Poisonous Mushrooms of the Northern United States and Canada*. Minneapolis: University of Minnesota Press, 1985.

Arora, David. *All That the Rain Promises, and More . . .* Berkeley, Calif.: Ten Speed Press, 1991.

————. *Mushrooms Demystified*. Berkeley, Calif.: Ten Speed Press, 1986.

Fischer, David W., and Alan E. Bessette. *Edible Wild Mushrooms of North America*. Austin: University of Texas Press, 1992.

Freedman, Louise. *Wild About Mushrooms*. Berkeley, Calif.: Aris, 1987.

Holsten, Edward H., Paul E. Hennon, and Richard A. Werner. *Insects and Diseases of Alaskan Forests*. U.S.D.A. Forest Service, Alaska Region, Report Number 181, 1985.

Krieger, Louis C. C. *The Mushroom Handbook*. New York: Dover, 1967.

Largent, David L. *How to Identify Mushrooms to Genus I: Macroscopic Features*. Eureka, Calif.: Mad River Press, 1986.

Lincoff, Gary H. *The Audubon Society Field Guide to North American Mushrooms*. New York: Alfred Knopf, 1981.

McKenny, Margaret, and Daniel E. Stuntz, revised by Joseph F. Ammirati. *The New Savory Wild Mushroom*. Seattle: University of Washington Press, 1987.

McKnight, Kent H., and Vera B. McKnight. *A Field Guide to Mushrooms of North America*. Peterson Field Guide Series. Boston: Houghton Mifflin, 1987.

Orr, Robert T., and Dorothy B. Orr. *Mushrooms of Western North America*. Berkeley: University of California Press, 1979.

Smith, Alexander H. *A Field Guide to Western Mushrooms*. Ann Arbor: University of Michigan Press, 1975.

Resources

Alaska Mycological Society, P.O. Box 2526, Homer, AK 99603-2526.

Mushrooms, The Journal of Wild Mushrooming, 861 Harold Street, Moscow, ID 83843.

Mycological Society of San Francisco, P.O. Box 882163, San Francisco, CA 94188-2163.

North American Mycological Association, 3556 Oakwood, Ann Arbor, MI 48104-5213.

Puget Sound Mycological Association, Center for Urban Horticulture, GF-15, University of Washington, Seattle, WA 98195.

Index

Alaska Northwest Books™
*is proud to publish another volume in its
Alaska Pocket Guide series, designed with
the curious traveler in mind. Ask for books
in this series at your favorite bookstore,
or contact* Alaska Northwest Books™.

ALASKA NORTHWEST BOOKS™
An imprint of
Graphic Arts Center Publishing Company
P.O. Box 10306
Portland, OR 97210
800-452-3032